# YOUR KNOWLEDGE HAS VALUE

- We will publish your bachelor's and master's thesis, essays and papers

- Your own eBook and book - sold worldwide in all relevant shops

- Earn money with each sale

Upload your text at www.GRIN.com and publish for free

**Bibliographic information published by the German National Library:**

The German National Library lists this publication in the National Bibliography; detailed bibliographic data are available on the Internet at http://dnb.dnb.de .



**Imprint:**

Copyright © 2018 GRIN Verlag
Print and binding: Books on Demand GmbH, Norderstedt Germany
ISBN: 9783668837751

**This book at GRIN:**

https://www.grin.com/document/446272

**Jana Robertson**

# Effects of humans interfering in nature exemplified by the living conditions of whales

GRIN Verlag

**Bundesgymnasium und Bundesrealgymnasium Kufstein**

A-6330 Kufstein, Maderspergerstraße 3

Vorwissenschaftliche Arbeit
Maturajahrgang 2017/18

# Effects of humans interfering in nature exemplified by the living conditions of whales

Verfasser/in:     Jana Robertson

Klasse:     8AIL

Einreichdatum:     *23.2.2018*

# Abstract

Our deserts are changing, our forests are changing, our oceans are changing, our environment is changing – our world is changing. And we are the major contributing factor. The human impact on the environment has risen to an immense level over the last few decades. Through our behaviour and daily actions, we cause global warming, the destruction of ecosystems around the world, pollution, deforestation and mass extinction. Human activities are forcing whales to leave their natural habitats and to face threats every single day. Therefore, I have chosen to focus my paper on the topic "Effects of humans interfering in nature exemplified by the living conditions of whales". The following chapters will deal with the dangers that whales face in the modern world – overfishing, contamination and pollution of the seas, noise influence and the threat of being caught and held captive in a marine park for example. Furthermore, this work will examine the importance of supporting the protection and preservation of the marine environment and the whales living in it through intensifying whale hunting regulations and through individual and collective efforts.

The results I obtained throughout my research sadly confirmed my suspicions: We have reached the point of no return. Several different marine species have already vanished from our planet because of human interference in their environment. Although many organisations have tried, and are still desperately trying, to save the marine environment they have, unfortunately, failed repeatedly. Immediate action is essential! This affects us all. Earth is the only planet we have.

# Table of contents

# 1  Introduction

"There's enough on this planet for everyone's needs but not for everyone's greed." –Mahatma Gandhi. This is a quote that accompanied me throughout my research and writing during which the cruel truth that lies behind it became clear.

When I chose to dedicate my work to this subject – one that is of great importance to me – the following questions presented themselves as most relevant: What are the main reasons for whales being threatened to the point of extinction? How severely does pollution, and the current problem of plastic rubbish, that is spreading to an unimaginable level, endanger whales? What impact does captivity have on the behaviour of whales? And what can we do to end whale hunting and the cruelty that these animals must face daily?


Every day we hear about environmental issues on the news, we read about animal species becoming extinct, we see photographs of Greenpeace activists and so on and so forth. However, we cannot allow ourselves to become desensitised through all the media coverage. The marine environment, our oceans, its inhabitants, including whales, are important for the preservation of our all lives. Without life in our seas we would not be able to survive.

Since this topic has always been an issue of great importance to me I had no difficulty finding suitable data through (mostly) non-governmental websites. Moreover, to obtain further information on what each and every individual can actively do to help save the whales, and to find out what specific organisations do in their efforts to protect marine mammals, I conducted an information exchange via email with Jane Wheeler, a member of Whale and Dolphin Conservation (WDC). Furthermore, my research into this topic enabled me to better understand the vital importance of whale conservation. Writing this paper offers me a way to pass on to others what I learned during my research and to perhaps convince people of the necessity to protect our marine environment and its inhabitants.

The aim of this paper is, on the one hand, to show how modern-day human behaviour is affecting the lives of many other species on planet earth – especially in our oceans. On the other hand, I intend to affirm that we can still reduce the dangers faced by marine mammals and, even with the smallest of efforts, everyone can contribute to the rescue of whales. My goal, furthermore, was to attain more personal knowledge on the topic of saving our planet and on saving whales living in our oceans, in particular. In this aspect I succeeded to my full satisfaction.

I divided my work into two main subjects: Firstly, how humans affect nature using the example of whales and, secondly, measures previously and currently taken to protect whales. I subdivided the first chapter into the following areas: threats caused by overfishing, contamination and pollution, subaqueous noise and changes in behaviour in whales held in captivity. The second chapter has been structured into three sections: intensification of regulations relating to whaling, measures that every individual can undertake to support organisations working towards the rescue of the marine environment and, finally, what these organisations actually do to help.

# 2 Effects of humans interfering in nature exemplified by the living conditions of whales

For decades now, it has been obvious that everyday human activities have been resulting in pollution, global warming, deforestation, overpopulation, excessive waste production, mutations of animals and plants and many other environmental problems. Human interference in nature causes severe dangers and threats to animals and plants all over the planet. We humans are even harming ourselves. This chapter aims to inform about the various ways humans change the living conditions of whales by interfering in nature.

## 2.1 Overfishing

The ocean has always been, and still is, the largest food source on planet earth and its fish makes up the main daily source of protein for 1.2 billion people worldwide. But in the last 60 years stocks have decreased by about 90% and scientists claim that the collapse of the oceans' entire fish population could occur in less than 50 years (cf. youtube.com #ending-overfishing TC 0:16-0:50). The reason for this is overfishing: "Overfishing occurs when more fish are caught than the population can replace through natural reproduction", as defined by the World Wildlife Fund (WWF) (worldwildlife.org #threats #overfishing). Capturing, processing and selling as many fish as possible may sound like a highly profitable business, but overfishing is a topic which should not be treated lightly, especially nowadays where fish stocks of several kinds are declining drastically. Overfishing is threatening the survival of marine mammals in two ways. Firstly, there is the danger of them becoming trapped in the nets as bycatch and, secondly, whales and other ocean-living mammals are facing a massive decrease in their food supply.

The beginnings of overfishing date back to the early 1800s when some species of fish, such as the Atlantic cod and herring and Californian sardines were caught almost to the point of extinction by around 1900. In the mid-20th century the demand for food rich in protein was incredibly high and governments made a huge effort to expand the fishing industry. Very soon consumers became accustomed to having easy access to a wide variety of fish species at very moderate prices. The fishing industry reached its climax in 1989 when approximately 90 million tons of haul were documented to have been caught only during this one year (cf. nationalgeographic.com #environment #oceans #overfishing).

One of the causes of this problem lies in "open access fisheries", i.e. areas in which the right to catch fish is free and open to all. This results in extreme overfishing and low profits for the fishermen. There are almost no property rights in the open sea and, such being the case, traceability of fishing activities is highly complex. But still there is no doubt that the main reason for overfishing is lack of control and government regulations. This makes it tremendously difficult for customs agencies and retailers to keep control of which fish are being imported into their country and, thus, they cannot guarantee that fish being sold have been legally and sustainably captured. Illegal fishing represents about 20% of the world's catches and can reach almost 50% in certain individual fisheries. This criminal act costs up to $23.5 billion annually. Moreover, fishing fleets are far larger than those required to catch the permitted amount of fish. The quantity of haul that has been gathered over the past centuries would cover an area comparable to four times that of the earth (cf. overfishing.org). Not only the constantly decreasing amount of fish in the ocean affects the living conditions of whales, as most species rely on fish as their main source of food, but also the increased shipping industry through fishing fleets is harming these marine mammals in a very dangerous way (as explained in chapter 2.3). A Toothed whales' diet (killer whale), as an example, consists of mostly squid, crabs and numerous species of fish. Through bycatch (elaborated later) not only too many fish are caught but also other oceanic creatures like these squid and crabs are to be overfished.

Some methods of catching fish are extremely counterproductive and severely harm marine ecosystems. The fishing industry is continuing to advance its hunting methods: its vessels are being adapted to go further, stay longer and catch more haul. Some ships are specially equipped to process and package the fish, with enormous freezing machines and powerful engines enabling them to carry the massive amounts of draught and equipment. They are, basically, giant "floating factories". Super-sized trawling vessels with lengths of up to 144 metres are big enough to hold more than 7,000 tons of fish. However, most of the haul is bycatch, i.e. fish that accidentally get caught inside the nets. Shrimp trawlers throw 80-90% of aforesaid bycatch back into the sea. This means that for 1kg of shrimps up to 9kg of other marine creatures accidentally caught are wasted. But not only do the affected fish populations – and therefore also the whole balance of the marine ecosystem – suffer because of these modern-day methods. Billions of lives of people who live in coastal areas all over the world, East Africa for example, and who depend on fish for their survival are also affected. These populations who rely on the available source fish suffer mainly from a lack of protein. For hundreds of years people believed that our seas would provide us with a limitless quantity of (sea)food. Modern fishing practices and increased demand over the last 50 years are causing the extinction of several species and we are slowly but surely heading to a point of no return, as various fish stocks important for daily trade – such as the Atlantic bluefin tuna – are under threat of vanishing completely (95% have already disappeared), while other smaller species – such as sardines and anchovies – are in danger of becoming too abundant. The World Wildlife Fund claims that "More than 85%of the world's fisheries have been pushed to or beyond their biological limits and are in need of strict management plans to restore them." (worldwildlife.org #threats #overfishing).

> "A new study conducted by the International Union for Conservation of Nature (IUCN) found that 5 out of the 8 tuna species are at risk of extinction. All three species of bluefin tuna, for example, are threatened with extinction and are at a population that makes their recovery practically irreversible." (eschooltoday.com #overfishing #impacts)

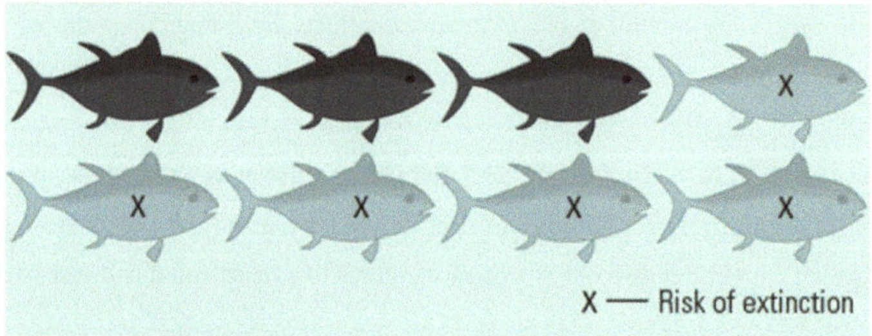


***Figure 1: Biodiversity***[1]

Almost 65% of worldwide fish populations today are said to be overfished and about 80% of all top predators are rapidly vanishing from the North Pacific and North Atlantic. European eels have almost completely disappeared from all oceans; salmon has left several rivers alongside the Atlantic and is now one of many threatened species, likewise sharks and rays. This could become a severe problem for killer whales in near future as they also rely on sharks as a part of their daily diet. The fact that 40,000 jobs were lost due to the disappearance of only one overfished cod stock in Canada in 1992 offers evidence of only one of several possible consequences of overfishing. And now, because the industry has led to depletion of fish in former hunting areas and fishermen are losing their labour, fishing fleets are seeking new untouched waters (cf. greenpeace.org #what-we-do #oceans #fit-for-future #overfishing).

---

[1] eschooltoday.com #overfishing #impacts

Another reason for the vast depletion of a huge number of fish stocks is "ghost fishing". This occurs when vessels lose their nets, something that quite commonly happens during a stay of several weeks or months. These nets do not only trap and kill fish but also some smaller whales like the sperm or beluga whale. They take numerous decades to degrade and the devastation they cause should not be overlooked (cf. eschooltoday.com #overfishing #impacts).

## 2.2 The danger of contamination of whales' living environment through toxic materials and plastic rubbish

In the oceans various kinds of pollution are to be found, such as rubbish, oil, chemicals and sewage. However, fact is that marine pollution is mainly caused by land-based activities. In the last decades our waters have had to bear the brunt of diverse, mostly serious, pollution accidents and although they are becoming less common because of improved modern technologies, there is still an unimaginable amount to learn about the damaging consequences of marine pollution (cf. ypte.org.uk #sea-pollution).

### 2.2.1 Plastic pollution

"Nearly all the plastic items in our lives begin as little manufactured pellets of raw plastic resin, which are known in the industry as nurdles. More than 100 billion kilograms of them are shipped around the world every year, delivered to processing plants and then heated up, treated with other chemicals, stretched and moulded into our familiar products, containers and packaging. During their loadings and unloadings, however, nurdles have a knack for spilling and escaping. They float wonderfully and can now be found in every ocean in the world, hence their new nickname: mermaid's tears." (telegraph.co.uk #drowning-plastic)

Approximately 6 million tons of rubbish enter the sea every year. Almost everything we dispose of incorrectly finds its way into the ocean. Since plastic rubbish often gets mistaken for food by marine mammals, countless inhabitants have been found choked to death through plastic bags and six-pack rings blocking breathing passages. This problem affects several marine species such as whales, dolphins, seals, puffins and turtles. Sperm whales are known to eat 100 million of tons of seafood each year, but nowadays their diet contains our waste too. The rubbish the marine animals consume takes up a lot of space which means that they lack important nutrients they should be taking in instead. In addition, some plastic items with sharp edges can slice open animals' throats and internal organs and large pieces may get stuck in breathing passages or digestive tracts.

Microplastics are found in nearly every single area of sea water around the whole world. These particles are so tiny they are not even visible to the naked eye and they mainly enter the ocean through our washing machines. The three main fabric particles found were polyester, acrylic and nylon. These substances can be found in almost every piece of clothing and scientists have discovered that only one garment can release up to 1,900 of these microplastic particles per wash. Recent tests reveal that there are more microplastic particles in the water than plankton. Trillions of plastic items can be gathered into massive, swirling rubbish patches by rotating ocean currents, known as gyres, and then travel millions of kilometres just below the water surface. A "trash vortex" is the name given to such monstrous, swimming rubbish patches, one of which in the North Pacific is around the size of the state of Texas.

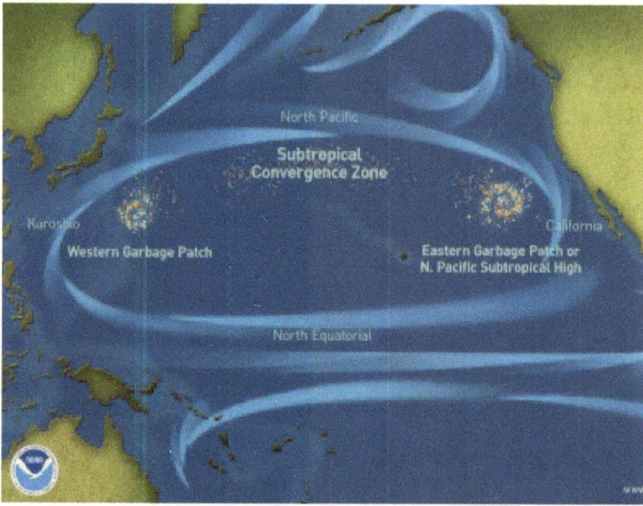


*Figure 2: Great Pacific Garbage Patch*[2]

Another patch was discovered in the Atlantic Ocean in 2010. The problem with plastic is that it never completely disappears from our seas. It does get smaller as the sunlight splits it up into tiny pieces over a long period of time through photodegradation, but this process allows toxic chemicals to spread in the oceans yet again (cf. ypte.org.uk #sea-pollution #plastic-pollution).

---

[2] marinedebris.noaa.gov #garbage-patch

## 2.2.2   Oil pollution

Oil spills are generally thought to be a huge catastrophe for the marine environment, but in fact they are responsible for only 12% of overall oil damage. Far more important is the industrial contribution which represents 36% of oil entering the seas through the same drains and rivers as chemicals do. Since the consistency of oil is very sticky, thick and slimy it can cause devastating damage to the marine eco system (cf. wwf.panda.org #oceans #problems #pollution).

The biggest disaster caused by an oil spill was the so-called "Deepwater Horizon oil spill" that occurred in the Gulf of Mexico in April 2010 where approximately 4 million barrels of oil leaked into the ocean and destroyed 4,000 miles of coastline. It is much harder to clear sandy beaches from the spilt oil than rocky cliffs and birds, especially, tend to get themselves completely covered in the sticky oil, making flying and swimming utterly impossible. Furthermore, marine mammals often swallow the slimy substance when trying to escape the mass on the surface and breathing passages get blocked, or internal organs are damaged which usually ends up in an excruciating death as crude oil includes over 1,000 chemicals of which some are extremely toxic. In warm waters the chemicals don't take as long to evaporate as in cold waters – here the light hydrocarbons on the water surface are often eliminated with fire (cf. ypte.org.uk #sea-pollution #oil-pollution). The website of National Oceanic and Atmospheric Administration (NOAA) contains the following statement: "We've learned from past experience with the 1989 Exxon Valdez oil spill that killer whales and other marine mammals don't avoid oiled areas on their own and exposure to oil likely can affect their populations." (response.restoration.noaa.gov #keep-killer-whales-away-oil-spill).

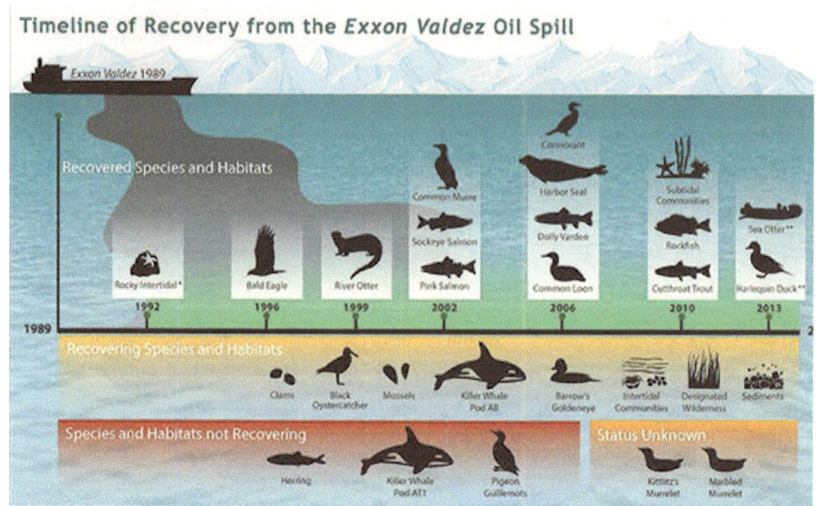


*Figure 3: Recovery timeline from the Exxon Valdez*[3]

From the figure above, we can see that a specific killer whale pod (AB) took 13 years to finally start recovering from the catastrophic oil spill, while other pods like the AT1 have never recovered again within the timeline of 1989 to 2014. (Pods within this figure were named by NOAA and U. S. Geological Survey (USGS).)

### 2.2.3 Chemical pollution

From the tiniest plankton to the largest blue whale – nearly every marine animal is affected by man-made chemicals. Until the 1970s almost everything was disposed of into the sea – even radioactive waste. In 1972 the London Dumping Convention imposed a ban on the disposal of the most toxic materials into the sea. In 1996 further restrictions on dumping were established.

However, today more than 100,000 commercially used chemicals and toxic fluids are still entering the oceans: via atmospheric transport, through discharge into rivers and streams or from direct disposal into the ocean. The main culprit is industry. A lot of factories ignore the

---

[3] oceanservice.noaa.gov #exxon-valdez

fact that their waste ends up in the sea and a few manufactures even dump their industrial waste at sea straight away.

Many pollutants enter the marine food chain through rivers and streams where the concentrations increase eventually reaching levels toxic for marine organisms.

Despite the severe consequences of chemical pollution and the fact that numerous marine species are threatened to the point of extinction, many humans are still unconcerned about this subject. In 1956 in Minamata Bay in Japan over a thousand people died due to the poisons that built up in fish and marine products they consumed. A specific factory was discharging small amounts of methyl mercury into the sea, but even this low concentration led to an enormous amount of suffering in the Japanese population (cf. ypte.org.uk #sea-pollution #chemical-pollution). The effects of the mercury poisoning are still being felt today. Some chemicals, when consumed, can lead to cancer, damage to the immune system, behavioural problems and limited fertility (cf. wwf.panda.org #oceans #problems #pollution). According to the website of the World Wide Fund for Nature (WWF), "Animals higher up the food chain, such as seals, can have contamination levels millions of times higher than the water in which they live. And polar bears, which feed on seals, can have contamination levels up to 3 billion times higher than their environment." (wwf.panda.org #oceans #problems #pollution).

It is almost unimaginable that people once thought the ocean would be big enough to "swallow" the amount of waste produced and dilute it to a safe level. They were so sure of it that they even came up with a catchphrase: "The solution to pollution is dilution." (cf. nationalgeographic.com #environment #oceans #pollution).

> "Today, we need look no further than the New Jersey-size dead zone that forms each summer in the Mississippi River Delta, or the thousand-mile-wide swath of decomposing plastic in the northern Pacific Ocean to see that this "dilution" policy has helped place a once flourishing ocean ecosystem on the brink of collapse. Scientists have counted some 400 such dead zones around the world." (nationalgeographic.com #environment #oceans #pollution)

Farmers spray toxic chemicals, such as pesticides and herbicides, onto their crops to improve quality and yield by controlling insect contamination and weed growth. As the toxic fluid finds its way into the soil after being applied to the plants it then enters streams and rivers and ends up in the sea. The chemicals mostly contain synthetic substances that do not biodegrade properly and appear to be resistant to natural breakdown processes (cf. ypte.org.uk #sea-pollution #chemical-pollution).

"The extra nutrients cause eutrophication - flourishing of algal blooms that deplete the water's dissolved oxygen and suffocate other marine life. Eutrophication has created enormous dead zones in several parts of the world, including the Gulf of Mexico and the Baltic Sea." (wwf.panda.org #oceans #problems #pollution)

Medication taken by humans (and animals) also contains harmful substances which make their way into marine environments through – in the case of humans – the sewage that we flush down our toilets. About 80% of urban sewage released into the Mediterranean Sea is untreated. The medication doses may be tolerable for us humans and some animals, but for many marine animals they can cause severe issues such as fertility problems in fish (cf. wwf.panda.org #oceans #problems #pollution).

At home, too, almost every individual flushes chemicals, and often toxic substances down the drain when cleaning, washing or polishing products. Harmful substances contained therein, such as sodium hypochlorite, petroleum distillates, phenol and cresol, ammonia and formaldehyde then enter the marine ecosystem (cf. ypte.org.uk #sea-pollution #chemical-pollution).

### 2.2.4    Sewage pollution

Several years ago, when people believed that the ocean could deal with a certain amount of sewage and that the substances would be diluted, many sewage pipelines were built. Today most of these have been blocked or removed but still there are some areas – mainly around the UK – where untreated sewage is released into the ocean every day. Also, gigantic cruise ships dump enormous amounts of sewage straight into the sea. "It is estimated that 95,000 cubic metres of sewage from toilets and 5,420,000 cubic metres of sewage from sinks, galleys and showers are released into the oceans each day.", claims Young People's Trust For the Environment (ypte.org.uk #sea-pollution #sewage).

### 2.2.5    Radioactive pollution

Nuclear waste, which is constantly increasing, is mainly produced at power stations, processing plants and military facilities. Since way back in 1952 small quantities of radioactive waste have been released into the Irish Sea, the English Channel and the Arctic Ocean. Radioactive waste must be kept in heavily insulated containers made of glass and concrete to ensure that the extremely toxic substances do not leak out. But through irresponsible or illegal dumping, which commonly occurs, radiation may enter the food chain via plankton or seaweed which are then eaten by smaller fish, then bigger aquatic animals until reaching the biggest of all – the blue whale. Seals and porpoises in the Irish Sea have already been found dead and poisoned by radioactive caesium and plutonium. The tragic catastrophe of Fukushima (2011) led to thousands of tons of radioactive waste being released into the Pacific Ocean and poisoning countless marine species (ypte.org.uk #sea-pollution #radioactive-waste).

## 2.3  Subaqueous noise

When you think of the word 'pollution' one of the first things that springs to mind is rubbish – especially plastic bags – and oil issues. But pollution isn't always meant in a physical or materialistic way. The oceans hold a vast amount of water enabling sound waves to travel freely for hundreds of miles. Noise levels caused by human activities have doubled every decade over the past 60 years. The increase in high-noise ships, sonar devices and oil platforms is also contributing to the damage of the marine environment and has a very negative impact on marine life. Every contributing component affect the migration, communication, hunting, protection and reproduction patterns of several marine species, especially whales. The oceans around Antarctica are a particularly important gathering and breeding spot for almost all cetaceans as these waters are – at the moment – one of the least affected areas with regard to noise pollution. Here the sound impact of the shipping industry is 20 decibels quieter than anywhere else on planet Earth as most ships are not equipped for the journey through the icy cold environment of the Antarctic Sea and furthermore, there is no oil exploration or production here. The Antarctic Peninsula, however, is frequently visited by tourists and researchers by ship, especially during the summer period. According to the Umweltbundesamt (Germany's main environmental protection agency), "There were at least 95 tourist vessels, 40 research vessels and 46 fishing vessels in the waters around the Antarctic during the 2012/2013 season. In addition, there are yachts and a few illegal fishing boats, most of which make several journeys per season." (cf. umweltbundesamt.de #international #antarctic #underwater-noise).

Although loud noises can also be used to guide whales, which communicate acoustically, away from danger zones to safe waters, the problem of underwater noise is still a serious

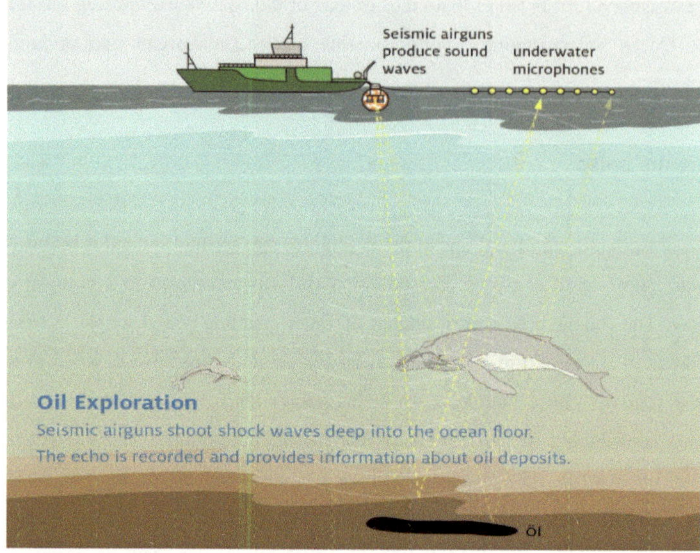


*Figure 4: Air guns*[4]

negative factor contributing to the decrease of whale populations in our oceans.

> "Seismic airguns are primarily used for oil and gas exploration below the seabed and for geophysical surveys of the seafloor. Air is driven into the water at high pressure, sending intense and explosive sound pulses towards the seabed. These sounds can permeate thousands of metres of ocean before penetrating hundreds of kilometres into the ocean floor. Up to 40 airguns are fired in a tight sequence, each of them emitting sound every ten to fifteen seconds, often for 24 hours a day and for several weeks in the same area. Hydrophones are used to record and analyse the sound that reflects back to the sea surface. As easily extractable resources are already depleted, seismic surveys keep spreading into ever more sensitive marine habitats and into ever greater depths." (oceancare.org #ocean-conservation #causes-underwater-noise)

These airguns are 1000 times louder than the noise a ship usually emits and the sound range of about 300 Hz is very similar to the main sound range of communicating baleen whales, such as the blue and humpback whale, which can frequently be observed in the Southern Ocean. The noise of underwater airguns is so intense that the hearing system of acoustic marine animals is often damaged. This acoustic trauma can finally lead to the complete loss of specific hearing abilities which are extremely necessary for survival.

---

[4] oceancare.org #ocean-conservation #causes-underwater-noise

The effects of underwater noise on marine mammals depend on the intensity, length and setting of sound exposure (sonication) (cf. oceancare.org #ocean-conservation #causes-underwater-noise).

The Umweltbundesamt (UBA) is currently working on an international sound protection concept specifically for the 20 marine mammal species native to waters around the Antarctic. A few whale species have learnt how to deal with the serious impact of noise pollution over the last decade while others are becoming increasingly distracted and their chance of survival is threatened:

> "Many whales will leave an area if it is too loud (flight reflex); others will become louder in a louder environment while still others cease to communicate entirely (altered communication behaviour). Marine noise can distract animals in such a way that they detect prey or a dangerous predator either too late or not at all (acoustic masking)." (umweltbundesamt.de #international #antarctic #underwater-noise)

Military sonar is used in exercises and training, often for detecting enemy submarines. These sonar systems emit strong sound pulses in all directions for hours without a break. Some sonar waves can fill up to thousands of cubic kilometres of ocean water with a penetrating sound of up to 230 decibels.

The use of underwater explosions to test military equipment also has a severe impact on marine life. The noise created by these detonations is immense.

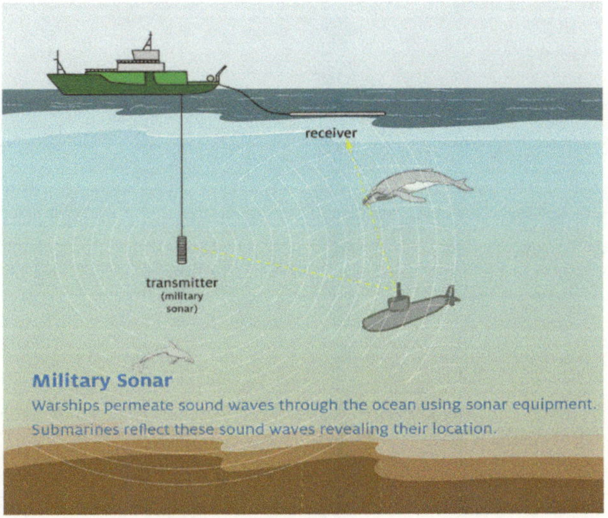


*Figure 5: Military sonar*[5]

---

Of course, the shipping industry is another major contributing factor, in fact it is the principal source, of underwater noise which is increasingly harming the marine environment. Today approximately 90% of transport is carried out by ships which produce a so called "acoustic fog" which, in actual fact, sounds very similar to natural underwater sounds such as those released by whales for example. Construction works in harbours, pile-driving for wind farms and oilrigs are also contributing to the immense noise emission in the ocean (cf. oceancare.org #ocean-conservation #causes-underwater-noise).

## 2.4   Changes in behaviour in whales held in captivity

There is one species of whale that is known to all as the most typical entertainment whale: the killer whale, also known as orca. This specific type of whale was introduced into the entertainment business in 1961 when people began to capture young animals and transport them to marine parks and zoos where they were put into tanks that were much too small and taught tricks and skills which they performed in front of big audiences.

As can be discerned from the statistics below in many countries the majority of marine mammals (especially orcas) in marine parks were born in captivity. In the United States, for example, only 25 out of 529 captive animals were born in their natural habitat.

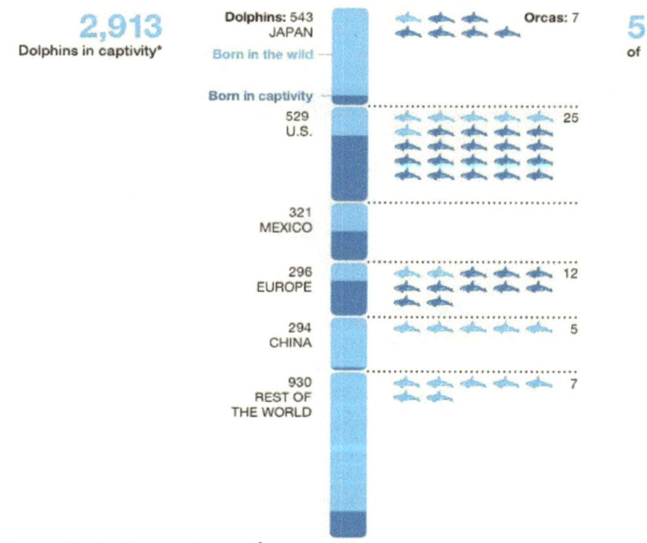


*Figure 6: Captive performers*[6]

---

[6] ngm.nationalgeographic.com #captive-performers

With its big marine parks Canada was once a major contributor to the display industry but today only one orca named Kiska remains in captivity. Orcas are also held in European marine parks (France, Spain and Russia) and wild killer whales are still hunted and captured in Russian waters. Some are even sold and shipped to China. Home to several captive orcas, the world's most popular marine park, SeaWorld, USA has finally stopped taking in orcas captured in the wild, but instead they are currently breeding them in captivity.

The reason for the popularity of this specific whale species is to be found in the nature and behaviour of the animal: Orcas are extremely friendly and learn quickly which makes working with them very easy. Furthermore, the life span of a killer whale and a human is more or less the same, so the bond between worker and animal – which is very important in this case – can grow stronger and develop over many years. The communication between the animals within one pod can also be compared to how humans communicate with each other. These animals have a highly complex brain. They demonstrate incredible intelligence and have even developed the ability to use a part of the brain we humans do not have. This ability brings the intensity of social bonding to a whole new level. Orcas are known to be the most social mammals on planet Earth – even more than humans. These animals live in "multi-generational units" and parents and male offspring whales within the pod stay together for their entire life. The young male may leave his group for mating for some time, but he will always return to his mother. Investigations show that killer whales are the only mammals that show this pattern of behaviour. Since 1961, when orcas first became popular in the entertainment industry, 165 orcas have died in marine parks all over the world (this figure does not include the large amount of miscarriages (30) and still-births) and 48 of these animals' lives came to an end at SeaWorld. Furthermore, out of 145 animals used for the entertainment industry (captured from the wild) only 20 are still alive (cf. us.whales.org #WDC-in-action #captive-orcas).

Living in a small tank deprives the animals of the chance to develop their survival skills and changes their natural behaviour completely. It is obvious that their natural physical demands cannot be fulfilled in a small tank with concrete walls surrounding it. The size of the pool an orca lives in when being held in captivity is so small that the marine mammal would have to encircle it over 1,400 times to match the distance it would normally cover in its natural habitat. Wild orcas tend to travel very long distances – up to 100 kilometres a day. A group of scientists once tagged a pod of orcas and found out that in one case this group of whales travelled for 42 days without stopping anywhere finally covering a distance of 9,400 kilometres. Large animals require habitats to suit their needs and this cannot be successfully achieved in a small artificial tank.

Furthermore, when young orcas are captured from a pod they are separated from their mother thus damaging a strong and very important bond that can lead to severe mental health issues. What is more, each pod of killer whales has its own unique call which they use for communication among group members. This behaviour, which is passed on from generation to generation, does not exist in small groups of whales living together in captivity.

In addition, in their own families orcas tend to hunt different species of prey. Whilst one may eat a specific type of salmon, another one may target seals and some orca groups even hunt smaller whale species like humpback whale calves. Up until today scientists have found ten different ecotypes (i.e. distinct units in which orcas live) which do not tend to breed with other ecotypes but only within one unit. In captivity this specialised hunting and breeding behaviour cannot develop properly. The captive orcas are not fed the food they would have evolved to prefer in a group of family members and they cannot breed with the same type of unit which makes every ecotype so unique. Different ecotypes do not normally interact with each other in the ocean so in captivity this can often lead to violent conflicts between two or more orcas when forced together.

Orcas show behavioural changes and mental health issues due to stressful conditions whilst living in a small tank. They frequently rub against tank walls, swim in circles repetitively, develop pecking behaviour and lie motionless on the water surface or on the ground of the tank – symptoms similar to prison neurosis. Some cetaceans are known to even slam themselves against the tank walls out of boredom and severe depression. A former SeaWorld orca trainer, John Hargrove, states that almost every killer whale vomited after a show. Records show that several cetaceans have tried leaping out of their tanks or eating large stones and many scientists now claim that this behaviour could be a desperate attempt at suicide. Some orcas receive addictive drugs similar to Valium to calm them down and prevent them from hurting themselves (cf. onegreenplanet.org #why-whales-and-dolphins-do-not-belong-in-tanks).

Moreover, scientists have now been above to prove that orcas held in captivity do not live as long as animals in the wild. Wild males can live up to 60 years and wild females even up to 100 years. According to SeaWorld, captive males usually live up to an age of 40 and females 50 years. But studies carried out in the past few years shows that the average age of captive orcas lies between only 6 and 12 years – the mortality rate is 2.5 times higher than that of wild killer wales and 92% do not survive past the age of 25 (cf. bbc.com #why-whales-should-not-be-kept-in-captivity).

One of the best-known symbols of the suffering of orcas living in captivity, and one that most people will have seen in films such as 'Free Willy', is the collapsed dorsal fin, also known as flaccid fin or folded fin syndrome. In the wild killer whales' fins are vertically stable and a male's fin can grow up to 6 feet in size. Since the fin of male orcas is taller they are more likely to collapse than those of female animals. Orcas in the wild travel long distances often at a high velocity and in deep water. The resulting "rushing movement" through the water provides the dorsal fin, which has no bones inside it, with enough pressure to keep the tissues inside healthy and straight. Since an orca in captivity spends most of its time on the surface of the water and does not have much opportunity for movement and exercise, the tissue in the fin does not get the support it would get in the ocean and it starts to collapse. The repetitive swimming-in-circles also contributes to this health condition. Another theory maintains that the dorsal fin collapses in captivity because of dehydration and the overheating of fin tissue as the water in the tanks in which they live is too warm. Folded fin syndrome occurs in less than 1% of wild orcas while those in captivity nearly all develop this condition (cf. bbc.com #why-whales-should-not-be-kept-in-captivity; onegreenplanet.org #why-whales-and-dolphins-do-not-belong-in-tanks). Incidentally, out of all the money that SeaWorld earns through their entertainment shows only a tiny amount is spent on or donated to killer whale conservation.

# 3  Measures taken to protect whales

"Whales and dolphins are remarkable. But why are they so important? Why do we need to end captivity, stop whaling, prevent deaths in fishing gear, and protect their homes, the oceans and rivers of the world? What is so special about whales and dolphins?" (us.whales.org #wdc-in-action #why-protect)

These are questions to which the Whale and Dolphin Conservation (WDC) team provides clear answers: Whales are essential for the health of the ocean ecosystem and, therefore, we need to ensure that their natural habitat is a safe place where they have a chance to survive, feed, breed, nurse and recover from catastrophes or human interferences in their living environment if necessary. It is very important to understand the causes behind the threats they face day to day so that we can effectively fight these and provide a healthy future for whales. This chapter aims to provide information on what we can do to help save and protect whales, on what has already been achieved with regard to their rescue and the elimination of danger and what is still to be done.

## 3.1  Intensification of regulations relating to whaling

Whaling is the act of hunting whales for their meat and also for bones and blubber which are processed into everyday products like transmission fluid, candles, margarine, jewellery, toys and tools. Several nations throughout the planet have been whaling for thousands of years. Hunting was first carried out way back in 6000 BC, mainly in the Arctic Ocean, the Atlantic and the Pacific by tribes such as the Inuit, Basque or Japanese to provide their people with food containing important vitamins, minerals and proteins. In addition to providing nutrients, the long-standing tradition was a part of their cultural identity. In the early days of whaling every possible part of the animal that could be used – including meat, skin, blubber (thick layer of fat under the skin of marine mammals), organs and baleen – were either eaten or processed, for example the baleen was made into woven baskets. As whaling continued and spread during the Middle Ages and Renaissance whale body parts were used not only for food and traditional goods but also for the fashion industry, particularly for corsets.

The whaling industry climaxed in the 18th and 19th centuries and quickly became a million-dollar business when whale oil was required more than ever, especially for machinery. In the 18th century fishermen were already experiencing difficulties spotting and capturing whales around the Atlantic coast. Therefore, the Americans expanded their whaling operations to the waters of the Arctic region. Their missions reached fever pitch in 1800 when modern technologies enabled effective hunting with minimal waste.

In the 1900s the Norwegians developed mechanised, steam-powered catcher boats with greatly improved weapons like deck cannons and heavy calibre harpoons. This technological progress allowed whale hunters to work more efficiently and even enabled capturing blue whales and fin whales which had previously been almost impossible to catch because of their speed and size. Until that point these particular whales could nearly only be taken and processed when stranded. And in the mid-20th century the Norwegian nation had already established multiple shore-whaling stations on over six continents. In the 1930s, despite government attempts to regulate hunting with international treaties, 50,000 whales were killed annually and by the end of the Second World War several species were devastated to the point of extinction. The USA finally made whaling illegal in 1971, but by that time 9 species were already claimed to be endangered and whale populations were constantly declining.


The dramatic rise in whale extinctions, however, awakened the interest of groups and organisations which tried to abolish or at least limit the killing of marine mammals. The International Whaling Commission (IWC), which was originally founded in 1946, slowly became an important contributing organisation which concentrated on monitoring whale populations. In 1972 a worldwide ban was finally established and in 1979 and 1994 multiple "whaling-free sanctuaries" were founded in the Indian Ocean and in some parts of the Antarctic. In 1982 the International Whaling Commission eventually issued a worldwide moratorium which completely prohibited the capturing and killing of whales. This ultimately took effect in 1987 and was accepted by most countries over the world (cf. nationalgeographic.org #history-whaling).

Unfortunately, Norway – one of the most notorious whaling countries – only accepted the ban until 1993 and then refused to end operations, especially on hunting minke whales. The Scandinavian country now sets its own quota on the number of whales it is allowed to kill. This number varies extremely from slaying 671 to over 1,000 minke whales.

"Norway is now hunting a higher proportion of breeding females which could put long-term survival of minke whales in the North Atlantic in severe danger.", as the International Fund for Animal Welfare claims (ifaw.org #which-countries-are-still-whaling).

Japan only agreed to the ban under pressure from the USA and has, since then, been fighting to withdraw its agreement and, in addition, has introduced its "scientific whaling programme". Even if nations, such as Japan, claim to only hunt specific species of marine mammals for scientific purposes there is evidence that the meat from whales killed in Norwegian and Japanese waters, for example, is marketed as food – often at very low prices in an attempt to expand consumption.



Likewise active in whaling is Iceland. The small island nation also established a so-called "scientific whaling programme" and in 1992 it completely withdrew from the IWC. In 2004 the country decided to re-enter the organisation but expressed its objections towards the moratorium and in 2006 Iceland started whaling again, capturing and killing a total of 208 endangered whales of various species.

Despite several failures IWC has made an important contribution to saving several species of whale and some stocks have already recovered from their depleted status. "The other thing that the IWC has very successfully done is to collect information and provide analysis of data to help us understand the status of various populations that in some cases we knew very little about.", says a research biologist at National Oceanic and Atmospheric Administration's (NOAA) Southwest Fisheries Center in California (nationalgeographic.org #history-whaling).

No agreement exists with regard to the hunting of small cetaceans as the international moratorium only applies to baleen and sperm whales. Analysis taken from DNA data over the past few years has identified 655 whale products marketed in Japan and Korea. Sei, fin, blue and humpback whales have been successfully protected and there have been no records of hunted animals of this kind since 1989. Furthermore, no export of whale products from Norway to Japan has been documented since 1991.

Although the general moratorium was successfully introduced in 1986 in most countries, limited whaling is permitted to indigenous groups of people native to a specific geographic area, for example in parts of Alaska, Greenland, Siberia, the USA and St. Vincent and the Grenadines (cf. onlinelibrary.wiley.com #wal-und-mensch).

## 3.2   Active help as an individual

Although a lot has already been achieved with regard to the protection of whales – the biggest step was the outlawing of commercial whaling with the implementation of the worldwide moratorium in 1987 – there is still an enormous amount of work to do before all countries on our planet stop killing, hunting, capturing and purchasing these marine mammals. Japan, Norway and Iceland, for instance, are still ignoring the moratorium and still refuse to give up their cultural tradition of hunting and capturing whales (cf. greenpeace.org #save-the-whales). About 3 million whales have been sentenced to death over the last century and not only whales but cetaceans of all kinds still face deadly threats each day. Some people are already working very actively to help save these majestic animals. And in helping save these mammals they are not only rescuing a marine species, but also mankind and the planet:

> "Researchers from the University of Vermont and Harvard University have determined that whales' feeding habits bring useable nutrients back to the sunlit surface waters where it can be used to "fertilize" phytoplankton, the very base of the marine food web and the source for at least half of the earth's oxygen." (onegreenplanet.org #five-things-you-can-do-to-help-whales)

An information exchange with Jane Wheeler – a member of the WDC organisation – via email on 5th January allowed me to understand the importance of our support for these vital organisations that are doing such an essential job helping not just the whales suffering from threats and dangers every day, but the whole environment, including us humans.

There are many ways in which individuals can help charities like WDC. One possibility is financial support and here it is very important to understand that even the smallest amount of money can help making a huge difference. A monthly donation of just 10 euros can help tremendously. Furthermore, Jane explained to me that donating a specific sum of money on a regular basis allows organisations to plan long-term campaigns and projects which are vital to successfully accomplishing their work. Single donations, however, may be more appropriate for people who do not live in the UK. All donations go towards helping to keep whales safe in their own habitat and to free them from unnatural living conditions if necessary (cf. uk.whales.org).

With WDC it is even possible to adopt an orca or a humpback whale. For only around 4 euros per month, you can give these highly intelligent animals a chance of living a safer life in near future. After you sign up for the adoption you receive detailed information about your adopted individual orca (cf. uk.whales.org #orca-adoption).

A lot of holiday operators offer so-called "swim with" experiences where tourists can touch and swim with whales. However, some travel operators, such as the WDC travel partner Alpha, have decided to exclude these acts as the "intrusion" of humans can be very stressful for the animals and scientists have found out that some cetaceans – when disturbed by humans – even leave their home area to find a quieter habitat to live in. The constant presence of humans wanting to interact while the animals feed, rest or nurse can have a long-term impact on the animals' behaviour and wellbeing. Last but not least, there is a high risk of marine animals becoming dependant on humans as they learn to accept food from them very easily (cf. uk.whales.org #lifestyle-giving).

Changing your everyday lifestyle a little can also help greatly. Recycling is something everyone can do to save the ocean and its whales. The more materials that are recycled, the less waste is created – waste that often ends up in the ocean. On a day to day basis individual consumers can avoid buying and/or consuming products that contain microplastics: A lot of cosmetics and skincare products (especially scrubs and exfoliation gels) here in Austria contain ingredients like polyethylene which belong to the group of microplastics. Furthermore, cutting down on unnecessary waste is another major contributing factor to help save the environment and its surroundings (cf. beatthemicrobead.org #product-table).

Raising awareness is probably the most important thing everyone can do to help, and this is the key to every organisation's success. Everyone can spread the word about issues close to their heart – in class, at work, in the newspaper or on the internet – sign petitions or write to influential figures who can do more and even represent the organisations and their work.


## 3.3　What whale protection organisations do based on the example of WDC – Whale and Dolphin Conservation

"Planet Earth needs healthy oceans to survive. And healthy oceans need whales. Whales also have an inherent value, which WDC believes should be recognised in law.", as Whale and Dolphin Conservation reminds (uk.whales.org #wdc-in-action).

In its attempt to rescue whales, the WDC organisation concentrates on three main topics: Firstly, WDC wants to see a total stop to the cruel act of whaling. Japan, Norway and Iceland are still catching and killing around 1,500 whales a year, despite the existing worldwide moratorium, the ban on trading in whale products and the constantly decreasing demand. Moreover, many of the animals die a slow and painful death, even if whalers claim otherwise. The organisation is fighting against whaling around the world, encouraging people not to eat or buy whale products and educating communities about the alternative of whale watching instead of "swim with" activities. To achieve its goal WDC is doing the following:

- "participating in international meetings where decisions are made about the future of whaling [...] and put the protection and conservation of all whales and dolphins on the top of governments agendas
- cutting supply routes [...]
- urging the EU to discuss whaling in any trade talks with Japan
- continuing to expose illegal sales of whale meet and have these operations shut down
- persuading tourists not to eat whale meat while on holiday [...]
- educating communities about the health risks of eating whale meat
- working to help local communities set up whale watching operations as a way of generating income rather than by killing whales" (uk.whales.org #wdc-in-action #stop-whaling)

Secondly, WDC wants to finally end whale captivity. It aims to free them from marine parks and the entertainment industry and rehabilitate these animals that belong in the ocean with their pod and not in small tanks. The organisation is currently working on this by creating sea sanctuaries where the animals that have been held in captivity for too long can live a more natural life as many, unfortunately, cannot return to their pod or even to the ocean. They are preventing the import of whales into marine parks, zoos and aquaria and at least reducing the demand for entertainment shows with whales by doing the following:

- "working with Merlin Entertainments on a project to rehabilitate and retire captive belugas to a natural sea pen
- providing expert advice on other potential sanctuaries around the globe
- stopping airlines transporting [...] to cut the supply of whales and dolphins to parks
- pressuring governments to ban the capture of wild whales and dolphins for display
- exposing welfare issues and cruelty associated with breeding in captivity
- campaigning to stop tour operators promoting trips to captivity shows
- working with local groups and individuals [...]" (uk.whales.org #wdc-in-action #end-captivity)

And, last but not least, WDC wants to finally stop whales dying in fishing nets as bycatch, since getting tangled in fishing gear is one of the biggest threats to whales and results in a slow and painful death. Just like humans, whales are not able to breathe underwater and the resulting panic they experience if caught in a net leads to severe wounds and broken bones when the animals struggle to escape. Some who do manage to escape carry parts of the nets around with them for years and eventually die from infections as the ropes cut into their flesh. When the marine mammals realise they cannot break free they close their blowhole and suffocate. The organisation is stopping whales dying as bycatch by doing the following:

• "campaigning for strong national UK laws to stop deaths in fishing gear [...]
• working within the EU to strengthen protection measures in Europe [...]
• arguing for better monitoring on-board fishing vessels
• representing whales and porpoises on federally appointed US task forces
• supporting scientists to find solutions and technologies [...] to help porpoises detect and avoid nets
• collaborating with fishermen to help them avoid entanglements
• petitioning the US and Canadian governments to protect the critically endangered North Atlantic right whales
• working in New Zealand to save the Maui and Hector's dolphins [...]" (uk.whales.org #wdc-in-action #end-bycatch)

# 4  Conclusion

Having reached the end of my paper I would like to reflect on the information I gathered during my research and writing process and, furthermore, pick up once more on the central questions posed in my introduction.

The main reason for whales being endangered and threatened is the interference in nature by humans. Through irresponsible activities and thoughtless, selfish behaviour we are polluting and contaminating the oceans and destroying whales' habitats. We are carelessly taking fish out of the seas without thinking about the consequences such as the death-by-bycatch problem, starving marine mammals and the extinction of whole species that lived on this planet long before the human race even existed. Plastic rubbish in the oceans is accumulating to such an immense level that entire plastic islands, so-called garbage patches, are forming and the number of marine mammals choking to death because they mistake the material for food is increasing constantly.

Not to be forgotten is the continuing practice of whaling which is still carried out in countries like Japan and Norway even though some species are barely surviving.

The popularity of killer whale shows in marine parks like SeaWorld is still a disturbing problem despite the desperate efforts of organisations and smaller groups to end captivity. The conditions under which orcas have to live in the tanks for years are devastating and even lead to acts comparable to suicide attempts.

Many people are unaware that there are many ways in which every single individual can help save the whales. Sometimes even the smallest effort can make a huge difference. We must learn to appreciate and respect the efforts of organisations working for the rescue of the marine ecosystem and its habitants and, most of all, we must understand the importance of our own contribution to saving planet earth.

Over the course of my work I found out that whales are, in actual fact, essential for a stable marine environment. Without them the oceans, and, thus, our planet and humanity cannot survive. This made me realise the importance of the existence of whales and, more crucially, the importance of the conservation of whales. Equally worthy of mention is the issue of already existing and possible future protection zones in which marine mammals can live without any danger or interference from humans. This is an area in which I would like to expand my knowledge further.

Through writing this paper I would like to draw this topic to the attention of others and make people aware of the fact that environmental protection, including the preservation of whales, is not a trend or a slogan, but is vital for the survival of our plant and of humanity.

# 5   Bibliography


Barre, Lynne: How do you keep killer whales away from an oil spill? 8.2.2016. https://response.restoration.noaa.gov/about/media/how-do-you-keep-killer-whales-away-oil-spill.html (4.11.2017)

Grant, Richard: Drowning in plastic. The Great Pacific Garbage patch is twice the size of France. 24.4.2009. http://www.telegraph.co.uk/news/earth/environment/5208645/Drowning-in-plastic-The-Great-Pacific-Garbage-Patch-is-twice-the-size-of-France.html (3.11.2017)

Hogenboom, Melissa: Why killer whales should not be kept in captivity. 10.3.2016. http://www.bbc.com/earth/story/20160310-why-killer-whales-should-not-be-kept-in-captivity (27.12.2017)

Holm, Patricia; Holm, Fynn; Holm, Jürgen: Jagen – beobachten – vergöttern. Wal und Mensch. 5.2016. http://onlinelibrary.wiley.com/doi/10.1002/biuz.201610602/epdf?r3_referer=wol&tracking_action=preview_click&show_checkout=1&purchase_referrer=www.google.at&purchase_site_license (8.9.2017)

Kennedy, Jennifer: Killer whale dorsal fin collapse. 28.4.2017. https://www.thoughtco.com/killer-whale-dorsal-fin-collapse-2291880 (27.12.2017)

Marrero, Meghan; Thornton, Stuart: Big fish. A brief history of whaling. 1.11.2011. https://www.nationalgeographic.org/news/big-fish-history-whaling/ (17.10.2017)

Umweltbundesamt: Underwater noise. 29.2.2016.

http://www.umweltbundesamt.de/en/underwater-noise#textpart-1 (6.11.2017)
http://www.greenpeace.org/international/en/campaigns/oceans/fit-for-the-future/overfishing/ (24.10.2017)
https://www.youtube.com/watch?v=F6nwZUkBeas (17.9.2017)
https://www.worldwildlife.org/threats/overfishing (19.10.2017)

https://www.nationalgeographic.com/environment/oceans/critical-issues-overfishing/ (19.10.2017)

http://overfishing.org/pages/what_is_overfishing.php (19.10.217)

http://www.eschooltoday.com/overfishing/impact-of-overfishing.html (24.10.2017)

http://www.greenpeace.org/international/en/campaigns/oceans/fit-for-the-future/overfishing/ (24.10.2017)

https://ypte.org.uk/factsheets/sea-pollution/polluting-the-seas?hide_donation_prompt=1 (26.10.2017)

https://ypte.org.uk/factsheets/sea-pollution/plastic-pollution (3.11.2017)

http://wwf.panda.org/about_our_earth/blue_planet/problems/pollution/ (2.11.2017)

https://ypte.org.uk/factsheets/sea-pollution/oil-pollution (3.11.2017)

https://ypte.org.uk/factsheets/sea-pollution/chemical-pollution#section (26.10.2017)

https://www.nationalgeographic.com/environment/oceans/critical-issues-marine-pollution/ (2.11.2017)

https://ypte.org.uk/factsheets/sea-pollution/sewage (3.11.2017)

https://ypte.org.uk/factsheets/sea-pollution/radioactive-waste (3.11.2017)

https://www.oceancare.org/en/our-work/ocean-conservation/underwater-noise/silent-oceans-causes-underwater-noise/ (4.11.2017)

http://us.whales.org/wdc-in-action/fate-of-captive-orcas (27.12.2017)

http://www.onegreenplanet.org/animalsandnature/why-whales-and-dolphins-do-not-belong-in-tanks/ (27.12.2017)

http://us.whales.org/wdc-in-action/why-protect-whales-and-dolphins-0 (13.1.2018)

http://www.ifaw.org/united-states/our-work/whales/which-countries-are-still-whaling (18.10.2017)

http://www.greenpeace.org/usa/oceans/save-the-whales/ (30.12.2017)

http://www.onegreenplanet.org/animalsandnature/five-things-you-can-do-to-help-whales-and-our-planet-right-now/ (30.12.2017)

http://uk.whales.org/support-us (5.1.2018)

http://uk.whales.org/adoptions/orcas (5.1.2018)

http://uk.whales.org/wdc-in-action/lifestyle-giving (5.1.2018)

http://www.beatthemicrobead.org/ProductTable.php?colour=2&country=OM&language=EN (7.1.2018)

http://uk.whales.org/wdc-in-action (7.1.2018)

http://uk.whales.org/wdc-in-action/stop-whaling-1 (7.1.2018) http://us.whales.org/wdc-in-action/end-captivity-1 (7.1.2018)

http://us.whales.org/wdc-in-action/end-bycatch-stop-deaths-in-fishing-gear-0 (7.1.2018)

# 6   Table of figures



http://www.eschooltoday.com/overfishing/impact-of-overfishing.html_(10.1.2018)

https://marinedebris.noaa.gov/info/patch.html (10.1.2018)

https://oceanservice.noaa.gov/podcast/mar14/mw122-exxonvaldez.html (10.1.2018)

https://www.oceancare.org/en/our-work/ocean-conservation/underwater-noise/silent-oceans-causes-underwater-noise/ (10.1.2018)

http://ngm.nationalgeographic.com/2015/06/rewilding-orcas/captive-graphic?_ga=2.150559538.567575934.1515607615-407356507.1508440794 (10.1.2018)

# YOUR KNOWLEDGE HAS VALUE

- We will publish your bachelor's and
  master's thesis, essays and papers

- Your own eBook and book -
  sold worldwide in all relevant shops

- Earn money with each sale

Upload your text at www.GRIN.com
and publish for free